John Boynton

Food for starving cattle on the plains

John Boynton

Food for starving cattle on the plains

ISBN/EAN: 9783337201241

Printed in Europe, USA, Canada, Australia, Japan

Cover: Foto ©berggeist007 / pixelio.de

More available books at **www.hansebooks.com**

Food for Starving Cattle on the Plains.

A TREATISE

ON

MAIZE, CLOVER,

SILOS AND ENSILAGE,

BY

JOHN F. BOYNTON, M. D.

PRAIRIE FODDER $1.00 PER TON.

*"Oh, how I pity the starving,
Freezing Cattle on the Plains."*
—*Pullman Passenger.*

SYRACUSE, N. Y.:

COLUMBIA PRESS, HERALD BUILDING,

1884.

"AMERICAN, BOYNTON ASSOCIATION"

I Cheerfully Dedicate
                    The following pages :

They are the results of experiments and thoughts during leisure hours while I was lecturing in the years 1878 and 1879 on Geology, Agricultural Chemistry, Fertilization of "worn out farms" and exhausted soil. They were intended as serials for a periodical, written, laid aside and nearly forgotten, until I was induced by friends to publish in the present form.

On reading Monsieur Auguste Goffart's communication on " Silos and Ensilage," I became interested in the doings of several gentlemen who were among the first to construct Silos in America.

I watched the effects of this kind of fodder upon cattle which had subsisted on it for several months at a time, and my belief in its usefulness is stronger to-day than it was when I finished writing these pages more than three years ago.

Yours respectfully,

JOHN F. BOYNTON, M. D.

Highland Place, Syracuse, N. Y.
        August 14, 1884.

SILOS AND ENSILAGE.

What is a silo? The word Silo is from the Spanish, meaning a cell or pit, and as now used means a cell or pit for storing and preserving green food, for cattle, for the purpose of feeding to them in the winter season.

ENSILAGE.

What is ensilage? Food that has been put into this cell or pit, and preserved therein is en-cellated food and is called ensilage. The bee en-cells its honey; honey is therefore a natural encellated food, or ensilage for the bee. Some birds, also as the Carpenter bird of California, and some animals, as the squirrel encell (or store) their food, the former by burrowing in the thick bark of trees, the latter in holes and pits in the ground and in trees. These are all instances of encelling or storing food by nature's process.

The practice of encelling food for animals, though only recently introduced among our farmers, has come down to us, we have reason to believe, from the most ancient times. For instance, in the country of Moab, in the limestone rocks, immense numbers of cavities or cells are now to be seen, wherein the Moabites probably stored their grain, and other dry food, wherein they also encelled their GREEN food, the former to be used as food for men, and both as fodder for animals.

The modern practice of putting green food for cattle into a cell, or pit made into, or upon the ground comes from Southern Europe, and is most likely taken from the Syracusans (as the name Silo implies), and was undoubtedly first introduced into South America by missionaries, and thence into other portions of the Western Continent.

THE BUILDING OF THE SILO.

How to build a Silo : Select a place in high, dry ground, where an excavation can be made 10 or 15 feet in depth without reaching water, or being in danger from surface water. The width and length may be according to the convenience of the builder, or as may be required for the amount of green food to be stored for stock ; but the most convenient width for general use is from 12 to 14 or 16 feet, as lumber is usually sawed in these lengths, and a good proportionate length for a width of 12, or 14 feet, is 36 feet or more, say 48 the length, depending upon the quantity of food to be stored. Having made the excavation, wall it up with stones, or brick and mortar, or concrete may be used, being careful that the walls are vertical. Then plaster the walls with a rich mortar, making a smooth surface. The cell is now made.

Now take planks 8 to 12 inches in width, and length corresponding to the width of the cell or pit. These planks are to be used in covering and pressing down the green food in the pit, as will be explained. The pit or silo is now ready to be filled.

FILLING THE PIT OR SILO.

How to fill a silo: Go to the field and cut down the green corn, gather up and cart to the silo. These cornstalks or other green fodders are to be chopped or cut into lengths of one-half inch, or even shorter would be better; stalks, leaves and green ears being all cut together, thrown in to the silo and firmly trodden down as thrown in. Some persons have found this process of treading down the fodder most conveniently performed by a horse and two or three men or boys. Every farmer will know how to arrange a temporary gangway for letting the horse down into the pit in the first place. The best method of cutting, is to have a strong hay cutter run by one or two horse power (or portable steam engine), as is a threshing machine.

When the silo is full and compactly trodden down (it is not necessary that the pit should be absolutely full before covering), the planks are to be placed on the chopped corn, or other fodder. And I would advise that a thin batton three or four inches wide be placed under the joining of the planks, so that the air cannot pass in through the cracks. Having well covered all the surface of the fodder the planks should be weighted down with about 150 pounds to the square foot, this pressure being for the purpose of excluding all the air possible from the chopped material. Now as the planks settle down upon the contents of the silo, if the walls have not been made true and smooth, the planks will catch, and not only injure the walls, but leave a space for the air to come in contact with the ensilage, which should be avoided. The amount of settling and compressing of the contents will depend greatly upon the firmness with which it has been packed and the amount of weight placed upon it. One silo examined which was 14x36 feet, trodden by a horse and two boys, settled only six inches under the weight of nearly thirty tons of boulder stones. Where stones, or other heavy material are not to be had, clay, or any other heavy earth (not gritty) can be thrown as weight on top of the planks, and will be found very convenient, as it can be shoveled off and allowed to remain in side heaps for another season. Heavy earth, like clay, has also the advantage of securing tighter joints. Also as it becomes dry from one season to the next, this earth can be moistened either before or after being thrown on again, so that it will pack more firmly. Or, if found more convenient, bolts of logs three or four feet long can be used as weightage. Suggestions of this kind will present themselves to the mind of any thinking farmer. At this point I make one suggestion. Some persons place dry straw over the contents of the pit before placing on the planks, so that this straw comes between the fodder and the planks, thinking that by this means they best exclude the air. But the solid planks alone are best, as the planks lying next to the ensilage are better than any straw, for the straw is liable to become mouldy and to decay, as it holds more air, which promotes decomposition and thus effects the surface of the ensilrge beneath it. The great secret in filling the silo and preserving the contents consists simply in pressing out, as nearly as possible all the air, and by close covering, protecting the ensilage from air getting in and from all external atmospheric influences whatever. The silo being filled, pressed down and thoroughly covered, the contents are now allowed to remain for months, if necessary, or until wanted for use, when it may be opened in the following way:

Take off three or four planks at one end of the silo, and remove the ensilage as wanted until the bottom of the pit is reached, thus leaving the mass of ensilage with a clean cut wall. Continue to remove plank by plank, as the ensilage is needed, always cutting through from the top to the bottom, before removing another plank, and so on until the silo is emptied. As the planks are removed from time to time it would be well to set them up endways at the opened end of the pit.

Mr. Austin Avery, President of the Milk Association of Syracuse, N. Y., has his silo built on the side of a hill, beneath his barns, so that a door enters through the walled side near the end Having once cut down to the bottom, he can subsequently enter this cleared place by this doorway and thus the remaining ensilage, as it is taken down, always from the top, after removing a few planks, can be more conveniently taken out through this door. If found more convenient the ensilage can be raised from the silo in a bucket or basket, arranged on pulleys with a windlass; or an inclined passage way can be made leading from the outside to a door in the wall of the silo, in a way similar to that described in Mr. Avery's silo. If the silo is built outside of a building already erected it will be necessary to well roof it, so as to protect the contents from the weather during the process of filling, and afterwards.

Care must be taken that the eaves of the roof extend a good distance over the edges of the silo walls, leaving a margin of earth that will not become saturated with water, and it would be well to carry off the water from the roof by means of eaves spouts. This margin of dry earth can be further protected by grouting around the building, coal-tar asphalts (pitch) and gravel, such as is used for sidewalks, being the cheapest for this purpose, or if more convenient cement can be used. The object of keeping this margin of dry earth around the silo is to guard against water being absorbed by the soil, then freezing, expanding, and displacing the walls. As the roof of the silo is merely for protecting the contents from the rain and snow, it will be only necessary to construct it as we do the covers of our salt vats at Syracuse, which are so arranged that one man can run them on and off readily ; so that when filling the silo, or when wishing to take out any ensilage it may be opened and closed, or partly so, with a mere push of the hand. These roofs are permanent, and are secured with a hasp, so that the wind will not blow them off, and they keep the vats closed, and thoroughly protected from the weather. Of course, in this case the silo must be finished off with a plate for the roof to stand upon. A great part of the above precaution will be found unnecessary if the silo is built under some portion of a barn, or other suitable building already constructed. I would recommend the construction of a silo at one side of the barn floor so as to cart the corn directly on to the floor of the barn, where the cutter and power can be arranged in such a way that the chopped fodder may fall directly into the silo as fast as cut. Every farmer, or stock raiser will have ingenuity enough to devise means for carrying out details of this plan according to circumstances.

WILL THIS ENSILAGE ROT?

This is a question frequently asked : will or will not, this encellated food rot in this silo, and become unfit for fodder? It will not, if prepared at the right time and in the right way. The way has been stated above and should be strictly followed, and, if followed, the ensilage will keep good during the feeding season of the fall, winter and spring months.

It will also keep well in the Southern States during the shorter winters, in which they are obliged to feed their cattle. The time is now to be considered.

WHEN SHALL THE CORN BE CUT?

All animated nature reaches its perfection during the season of fecundation. The bird displays his most gorgeous plumage and sings his sweetest song at the time when he seeks his mate, and at this time, literally as well as figuratively, when the male or fertilizing principle is most active, the birds are in the full sweetness and perfection of their being. The mating season over, and the young birds hatched, the song of the male dies away, his plumage loses its brilliancy, and he returns to his normal condition. It is at this season, too, that the glow worm trims its lamp, and shines the brighter to attract its less brilliant mate. Even the cold-blooded and unimpassioned fish, as the season of reproducing its kind approaches, with quickened powers brightens its gilded scales, though obedient only to natural law. So with vegetable life. All the grasses, corn which is only a mammoth grass, plants, etc., contain their greatest amount of sweetness, and are in their utmost perfection at the fertilizing season when root and stalk and every tender blade is filled with its richest juices, the whole plant contains the greatest amount of nourishment, while the blossom when in its perfect state throws off its soft aroma to the air. All plants are in their sweetest and most vigorous condition when in blossom. They then hold the most nutriment, the tissues are more tender, more easily digested (having less woody fibre) are more palatable, and are eaten with a greater relish by the animal. At this time much of the cellulose of the stalks is in a partly gelatinous condition, and is easily acted upon by the weaker acids and is most readily digested and assimilated.

The corn stalks are in the most favorable condition to be used as ensilage when the pollen is just ripe, that is, when the spire is in full bloom and beginning to shed its pollen upon the silking ear below. At this time the corn stalk is filled to repletion with juices, has its full amount of sugar and starch, of albuminous, nitrogenous and hydro-carbonaceous (or fatty, buttery) compounds and proteine substances; it is now better adapted for mastication, to the digestive and assimilating functions of the animal economy, and is altogether in its perfected state for supporting animal life.

I have spoken of the silking ear. Yes; for the silk of the ear is but a tassel of vital tubes, their office being to absorb and carry the pollen's vital energy, or fertilizing principle to the forming, swelling germs in the ear; therefore, when the corn, putting on the adornment of its early green, its stalks and blade and ear swelling with the full flow of its perfect sweetness, and its silken threads have colored themselves with their daintiest gold, the whole plant has reached the stage of its vital maturity and its acme of nutritive powers as food for animals. Then, and not till then, the pollen trembling on top of the slender spire, is showered upon these drooping threads of silk, the fertilizing principle is carried down through these delicate tubes to the waiting germ below and each kernel is fertilized and vitalized and developed into the perfect grain.

From this point the plant now advances to its ligneous or woody condition, and its better qualities are now carried farward to ripen and plump the seed as a store house for the coming life, that the plant may continue its species.

There is a silken tube for every germ, for every kernel of corn that is to find its place upon the perfect ear ; and therefore should the wind blow some of the pollen dust away, for every tube that does not receive its portion of pollen or fructifying principle there will be an undeveloped germ and a blasted kernel ; although other causes, as injuries, etc., may prevent the development of the full vital power of the germ.

It is at this quickening season of fructification, therefore, when all animate nature is most active, that grasses, corn, and food nourishing plants generally for animals and man contain, as we have seen, the sum of all their nutritive qualities. When the pollen, then, is falling and the kernel is first in the milk, is the most favorable time for cutting and gathering corn stalks for the silo. An experienced farmer will know how to average the gathering of his corn, as some varieties of corn will drop their pollen earlier than others.

I am of the opinion that corn of the largest growth is the most advantageous for encelling, as it furnishes the most fodder.

Hence I would use Dent or Southern corn, as the Northern season will furnish sufficient time for the stalk to grow, and for the purpose of encelling we do not wait for the grain, but only for the pollen to ripen. In experiments made in grinding and crushing corn stalks for sugar making I have found that the stalk of one variety of corn contains about as much sugar (or sweetness) as another of the same weight, if cut at the right season; but the varieties of sweet corns have their special sweetness in the kernel, a portion of its starch having been converted into natural glucose.

In planting this corn I would never sow nearer than about six inches for the larger varieties of corn, and about four inches for the smaller, the drills being from three and one-half to four feet apart, depending upon the variety of corn used.

Georgia corn stalks are extremely succulent, full of sugar, starch, gelatinous, partly digestable cellulose. They sometimes grow to about sixteen feet in height and nearly seven inches in circumference, and, in favorable soils, have very long, broad leaves.

One ton of Southern corn stalks will yield twice as much fodder as Northern corn. Twelve green Southern corn stalks have been known to weigh forty pounds. Some Kansas corn shown at the Centennial was said to be eighteen feet high and over seven inches in circumference at the butt when green

If the people of Kansas and Nebraska would encell their green corn while in blossom there would be no cause of their stock starving in the winter.

Manure well, plough deep, and plant when the soil is in good condition. Do not sow the corn, but plant it, and do not plant too thick. I should recommend to use one-half to one bushel of Georgia, Dent or Southern seed corn to the acre, as the kernel of some of the Southern corn is as large as two or three of the ordinary Northern kernels.

As corn extracts from the soil a large quantity of potash, wood ashes will be found a valuable fertilizer, particularly on old lands. Use unleached ashes, ten to twelve bushels to the acre (if leached ashes are used then triple the quantity). The farmer will find this a good investment.

Use, then, good seed corn of such varieties as you think may be best adapted to your soil and climate, recollecting that but few weeks are required for the larger varieties . to reach their blossoming period. I repeat the caution to use only a good quality of seed corn. If you have it in the ear, shell off the tips and butts and feed them to your fowls and stock, reserving the sound central kernels of the ear for planting if you want healthy, even growing shoots and the corn to be of average size and growth; and rest assured that by following this plan, if the larger varieties of corn have been planted, you will be rewarded with a bountiful crop of from forty to eighty tons to the acre, when the nurture of the sunshine and the rain has not been withheld and the seasons in their regular round have been favorable.

Should any of the planted kernels of corn fail to germinate and come up, or be destroyed by the cut worm, crow, etc., an earlier corn can be planted (being first soaked) in these vacant places in the drills, and this earlier variety will reach the blossoming period at the same time as the corn of larger growth. The old rule is to "make hay while the sun shines," but the advantage (and it is a good one) of this process of encelling fodder is that the corn can be gathered, cut and encelled in cloudy and damp weather, when hay cannot be made, and thus the "coming farmer" can literally be making hay (or fodder) without the aid of sunshine.

SILOS AND ENSILAGE.

The Economy of Ensilage—Cost of Feeding—Yield —Production.   (Mr. J. M. Bailey.)

It is stated on good authority that the lowest cost at which a cow can be kept in Eastern Massachusetts is 22 cents per day for feed, allowing nothing for care except the manure.

This amounts to forty dollars and fifteen cents for one cow for six months. Now let us see what ensilage will do.

First let us find what it will cost to lay the corn fodder down in the silo, where it becomes ensilage.

In a country where corn grows thriftily and wages are one dollar and fifty cents per day (or average that), the corn fodder can be raised, cut, and stored in the silo for one dollar per ton (of ensilage) and this ton of ensilage, if good ensilage rightly prepared, is believed to be equal to more than one-half ton of "Timothy Herd's grass," hay, which sells at from ten dollars to twenty dollars per ton, according to the supply of the season and the price at any particular locality. One gentleman, on being asked, told me that it cost him about eighty cents to one dollar per ton for cutting the corn in the field, hauling it to the cutter, chopping it about half an inch in length and packing it in the silo; "much depending," he added, on horse feed, wages and board.

This may be a close calculation, but without doubt, taking the results of many practical experiments, a fair and even liberal general calculation for the cost of raising, gathering, hauling, cutting and laying the corn fodder in the silo, packed, pressed and covered, is not more than one mill per pound, or two dollars per ton. Upon this basis let us see what it will cost to keep one cow on ensilage alone for six months, the longest feeding season. One cubic foot of ensilage, well pressed, will weigh about fifty pounds. This is about sufficient for one days' rations for one cow. Now there are about forty cubic feet in one ton of well pressed ensilage, therefore one ton of ensilage will keep one cow forty days. Taking the calculation of two dollars per ton for the ensilage, the cost per day of keeping one cow on ensilage alone is as two dollars divided by forty, which is five cents per day; the cost of keeping one cow on ensilage for six months would therefore be nine dollars and twelve and one-half cents.

This is even a liberal calculation, for, as I shall show

further on, forty pounds of ensilage per day has been found sufficient to keep an animal in healthy condition when not working, giving milk, or carrying calf. Now if it take one cubic foot of ensilage per day for one cow it will require 182½, or, in round numbers, 180 cubic feet to keep her six months. The silo to hold this amount of ensilage would only need to be six feet long by three feet wide and by ten feet deep, which equals 180 cubic feet, the size required. Calculating 50 pounds of ensilage to the cubic foot—a fair average—this size silo will hold 4½ tons of ensilage.

These figures are, of course, subject to some variations, according to the compactness of the ensilage and the amount of water in it. Also the proportions here given for the silo can be varied from somewhat, so as to allow room for settling, for the planks and for the weightage for compression.

HOW MUCH ENSILAGE CAN BE GROWN TO THE ACRE?

It has been absolutely proved that 75 tons of corn fodder can be raised to the acre. If this quantity can be raised north, in the Eastern States, it is surely not too much to assert that 100 tons to the acre could be reached in our fertile prairies and in many portions of the Southern States. And if such quantities can be produced with our Northern corn, what results might we not look for when the large growing corns and sorghums are used in their own congenial climates in the place of maize as they are destined to be in the near future.

In some debates before the Onondaga Farmers' Club it was stated that "two tons of this ensilage are worth more to feed than four tons of common corn fodder, or than one ton of the best Timothy hay." Or, in other words, that one ton of ensilage is more than equal in feeding power to two tons of common corn fodder or to one-half ton of hay.

We have seen that 50 pounds of ensilage per day is rations for one cow. Now 50 pounds per day equals 9½ tons for the year for one cow. This clearly shows that one acre of Southern corn producing easily an average of 70 tons to the acre, will more than keep seven cows for one year.

It has been found that one cubic foot (50 pounds) of ensilage will keep one sheep one week.

It is admitted that from 40 to 75 tons of corn fodder or ensilage can be easily raised upon an acre, and is about equal to from 20 to 37½ tons of good hay ; or to state it more simply, one ton of ensilage is fully equal in feeding powers to one-half ton of the best hay. Or, to express it in dollars and cents, two dollars worth of ensilage is equal in feeding value to from five to ten dollars worth of hay, according to price of the latter.

The following statement and calculation shows the immense comparative value of corn fodder raised for ensilage over hay :

COMPARATIVE PROFIT OF ONE ACRE OF ENSILAGE OVER ONE ACRE OF TIMOTHY HAY.

We have shown that one acre can be made to produce 70 tons of fodder, which, laid down as ensilage, will feed seven cows for one year. One acre, in the same high state of cultivation can be made to produce three tons of Timothy hay. As it takes four tons of hay to feed one cow one year, this three tons of hay will feed ¾ of a cow for the same time; and the gain, therefore, in number of cows fed on the acre of ensilage is 6¼ cows.

The question is, what is the cost of the acre of ensilage over the acre of hay and what is the value of the product of the gain of 6¼ cows. The cost to the farmer of cutting and housing three tons of Timothy should not exceed five dollars; while the cost of 70 tons of ensilage, at the very high estimate of two dollars per ton, would be $140. We, therefore, charge the ensilage with the difference of $135, and credit the product of the 6¼ cows, which in butter and increase of stock could hardly be less than $40 for each cow, or $250. This shows as a gain in favor of ensilage of $115, as an offset to the extra labor of taking care of the 6¼ cows is the great increase in the manure. It is for the practical farmer to consider this point; that by means of ensilage he need take but a small portion of his farm for the feeding of his stock, leaving a large, and, if he will, the larger portion of his land for raising hay for baling or market use; and whether it is more to his profit to sell hay at three tons to the acre than to increase his stock and its products.

One thing is certain, that a man can raise on a small portion of land more feed for stock in this form of ensilage than in the form of hay, and, therefore, by means of silos and ensilage, he can triple or quadruple the number of his cattle, including sheep, hogs, etc., as a few acres manured and cultivated on correct scientific agricultural principles will yield larger proportionate results in stock than hundreds of acres cultivated as heretofore.

Practical and correct experiments with silos and ensilage have already proved beyond question that by means of ensilage larger herds of stock can be raised on the farm and that the stock thus raised will be in better and healthier condition than by the old ways, and at a less cost of labor, while netting better pecuniary results.

It must be borne in mind that all the figures and calculations given above about crops, yield and production, will depend greatly upon soil, climate, tillage, how well fertilized, the kind of corn planted, the variations of the seasons, and other natural causes affecting the growth of crops.

But if this ensilage will do only one-half of what we assert that it will do, it still proves that it is far superior to any mode of feeding or boarding stock that has yet been made known, and is certain to produce the most remarkable revolution that has ever taken place in agriculture.

Mr. J. M. Bailey, of Winning Farm, has proved by experiments upon bullocks that 40 pounds of ensilage, containing 80 per cent. of water, fed daily without any other food, was sufficient to sustain the functions of the animal in a healthy condition. If a gain in live weight is needed, however, or it is necessary to replace waste tissue as in cows giving milk, oxen or other draught animals, when working, or to replace accumulating tissue, as in cows carrying calves, then it becomes necessary to add other food, as bran, or other foods containing proteine and fattening substances.

During the period of gestation the cow must consume and appropriate an amount of food equal to the weight o the being she produces, or else there must be a draught upon her own physical structure. When a cow is giving milk her system must be more than merely sustained, as she is yielding a product of a certain number of pounds per day, according to the quantity and quality of milk she produces.

This amount of her product, then, must be re-supplied by an amount and kind of nourishment equal to the amount and quality of her production.

Now, will an increased quantity of ensilage alone, without other food, supply this demand? This is an ·important question which I am now having thoroughly tested by a number of practical experiments. If ensilage alone will do this, well and good; if not, then sprouts, bran, corn, shorts or other nitrogenous foods must be added in the required quantity.

By this system of preparing encellated food for cattle instead of the ordinary fodder, a great saving of both labor and space is gained.

The several processes of mowing, spreading, raking, cocking, haycapping, pitching, loading, unloading, stacking and stowing away in barns are all dispensed with, and the space thus saved can now be used for stabling and other more economical uses than mowing bulky hay which contains less palatable food, in proportion to the space occupied, than does the closely packed ensilage; and instead of the hay for stock this space can now be occupied by the stock itself, in increased numbers.

An important consideration in this system of encelling food is that the surplus or residue of many vegetable growths, to be found on every farm, can all, or nearly all, be made to contribute their portion to the silo. If you have any kind of grass, clover, timothy, millet or Hungarian grass *in bloom*, green oats, rye, pea vines, sugar beet leaves, or any other vegetables that cows relish, *and that will not give a bad aroma to the milk or butter*, cut and mix them with the chop as it goes into the silo.

The practical economy of this system of encelling fodder, as well as its adaptability to all circumstances, will perhaps be best understood by the following

ILLUSTRATION.

The poor woman with only one cow, can store up a large quantity of food for winter use in a very small space. She can improvise a silo, by taking some clean old casks or boxes, and setting them in the ground where they will keep cool, not freeze, and be free from water, or place them under a shed and fill them with the chopped produce of her little garden plot. A large quantity of green food can thus be encelled in a cask or other convenient receptacle, care being taken that it is properly weighted down, and the air *thoroughly excluded*. As fast as her corn is ready, she can chop it fine, put it into the cask, press it, and weight it down with stones (about 150 to 200 pounds) laid on the cover, which should be made a little smaller than the cask or box. A piece of India rubber cloth might be conveniently made use of to place over the chop before pressing down the head or settler, and by securing this cloth tight around the settling head, the air would be excluded, and the ensilage well preserved. This ensilage she can begin to use as soon as the fall pasturage gives out, care being taken to re-cover it closely each time any ensilage is removed.

He who has the boldness first to inaugurate among his own countrymen a new era, should have accorded to him the full credit resulting from his success, and therefore to Mr. J. M. Bailey, of Massachusetts, is due the honor of having been one of the first to make silos and ensilage a successful experiment in this country.

The following table by Mr. Bailey, comparing the expenses of a farm, cultivated in the old way *without* ensilage, and in the new way *with* ensilage, must be of interest to all farmers who have tried, or *ought* to try, silos and encelled fodder.

ESTIMATED AMOUNT OF FIFTY ACRE FARM, BY J. M. BAILEY OF WINNING FARM, FIFTEEN COWS, WITHOUT ENSILAGE.

EXPENDITURE.

Six per cent. interest on farm, value $5,000.........$300 00	
Repairs on buildings, 2½ per cent on $2,000......	50 00
Taxes on farm, $40; on stock, $10....................	50 00
Interest on stock and farming tools...................	90 00
Wages and board of hired man, 9 months, at $30..	270 00
Depreciation on stock and farming tools	
at 10 per cent	150 00
Total expense..............................,......$910 00	

Total income on 15 cows 2,000 quarts milk per	
cow, at 3 cents	900 00
Deficiency...	$10 00

SAME FARM WITH ENSILAGE, 28 COWS, 100 SHEEP, 7 HOGS.

EXPENDITURE.

Interest on farm, stock, silos, manure and sheep-
shed..$ 561 70
Wages, one hired man, 6 months, at $25.......... 150 00
Repairs...   50 00
Taxes and insurance...................................   80 00
Meal and bran, 4 pounds per cow, per day........ 280 00
Grain for sheep and horses............. .... 150 00

Total expense......................$1,271 70

INCOME.

5,600 pounds butter at 10 cents.....................$ 560 00
14,000 pounds pork at 3 cents....................... 420 00
28 yearlings, at $10................................. 280 00
700 pounds wool at 30 cents 210 00
90 lambs, (cotswold) at $4.......................... 360 00

Total income......................$1,830 00

Profit..............................$ 558 30

In a letter from Mr. August Goffart, of Burtin, France,
whose name will go down to posterity as one of the greatest benefactors of his race, he being the first to make successful experiments in silos and ensilage, written to Mr.
J. B Brown, President of the New York Plow Company.
He states:

"The longer experience I have in feeding ensilage to
stock, the more I am convinced of the great service it will
render to agriculture. From October, 1878, to October,
1879, I fed the hundred animals in my stable ex-

clusively with corn ensilage during the winter, and con-
currently with fresh maize (corn) at the time when I had
it. The animals have always enjoyed the most excellent
health, and I can assure you that they have more appetite for the ensilage than for the fresh fodder, whatever
kind it may be. In reckoning 6 per cent of the weight
of the animal for its daily food, I arrive at an expense of
3¾ of a cent per day to feed an animal of 1,320 pounds,
and the total cost of ensilage ready to be fed at 90 cents
per ton."

This low rate of cost given by Mr. Goffart may be easily
accounted for by the cheapness of labor in France, and by
the fact that the French people are more economical in
their general methods of work than we are in this country.

BENEFICIAL EFFECTS OF ENSILAGE UPON ANIMALS.

Ensilage is more relished by the cow than any food
that can be given in the winter season, and in every
known instance milch cows have increased the quantity
of milk when fed upon ensilage, the increase beginning
at once from the first day of feeding.

In the course of my investigations upon this subject, of
silos and ensilage, I have examined hundreds of
cattle fed upon ensilage and always with the following
results:—Their eyes were bright, their teeth clean and
white, breath sweet, skin soft, their hair smooth and sleek,
and, in fact, they all looked healthy. All the herds that
I have seen appeared contented and satisfied with their
food, devouring it with avidity. They would at all times
leave any other food when they could get ensilage. Mr.
Austin Avery, President of the Onondaga County Milk

Association, and a thrifty farmer, near Syracuse, N. Y., keeps 72 milch cows fed on ensilage with an equal quantity of dried corn stalks, cut and steamed. He told me his cows gave each one quart more milk when fed on ensilage than when fed on hay with shorts, meal or bran. Also that his milk, butter and cheese are as good, if not better, than when fed the old way. His horse is also fed on this ensilage and he is in better condition than when fed with hay and the usual quantity of provender and is doing his ordinary work.

In a visit paid by me to Mr. Van Orten's silo on his farm in Spring Valley, N. Y., he very kindly gave me the following information in regard to the effect of ensilage upon his animals and their products:

"All cows eat it with a relish. Three cows, newly purchased, did not eat it at first, as they were homesick, and gave but little milk, but when they became wonted to the place, and after they had eaten the ensilage, they came to an increase in their milk. Fresh cows, new to the ensilage, increased all of two quarts per day. Cows taken from grass and put on ensilage with some sprouts, meal, bran, etc., increased in milk over the grass feed." He fed about one bushel of ensilage to which he added, in some cases, two quarts of corn meal, in others four quarts of ground wheat middlings, to take the place of corn meal, (this being food for butter).

Some of his corn stalks were 13 feet high, had 16 leaves to the stalk, and were 1¾ inches in diameter at the ground.

Among other important advantages of ensilage is the fact that it does not physic or bloat animals.

LIVING DUST.

A!l hay, grains and dried vegetables, have more or less dust upon them, which on disturbing the hay, etc., rises and fills the air, is taken in by the breath and consumed by the cattle together with the hay and other vegetable fodder.

Now the larger proportion of this dust, is *living dust*, and is made up of microscopic germs of fungi and spores which excite fermentation in vegetable infusions. Nine-tenths of the dust arising from poorly cured hay, is found on miscroscopic examination to be composed of spores, which when fed to cattle in large quantities are liable to produce disease.

Rust, ergot and other fungoids on rye, smut on oats, wheat and corn, the various blights and destructive mildews on grasses and other vegetable growths; all these are but congregated masses of *disease-producing spores*.

Diseases of cattle, and even of man, are known to follow and correspond with the seasons when the various smuts and mildews of grains and other vegetable growths prevail; in other words, past observations have revealed the fact that the diseases of cattle follow the fungoid condition of their fodder.

Musty and mouldy hay that fills the barn with living dust, or the dust that may become alive when warmed by animal heat, and vivified by the juices of the body, bring about many of the diseases of the animal (perhaps even the untimely casting of the calf), the immediate causes of which are to be found on our own premises, and in our own neighborhood, and should not always be considered as epidemic, or as arising from atmospheric changes.

On examining by the microscope, these germs upon portions of plants that had been encelled, *their life* seemed to have been destroyed by the gases, vapors, and *acids* that had arisen during the period of ensillation.

Microscopic examinations have led me to the conclusion, that these fungi and spore forms of life, found so abundantly upon all vegetable growths, and which are so detrimental to the health of animals, have been destroyed by ensillation ; consequently, as this encelled food has been disinfected by the slight chemical changes it has undergone during its preservation in the Silo, I therefore feel confident in predicting, that cattle fed on good ensilage, will be either altogether exempt from many of the diseases which have hitherto afflicted them, or, that these particular sporoid diseases, themselves, will be greatly modified.

BENEFICIAL EFFECT OF ENSILAGE UPON ANIMAL PRODUCTS.

BUTTER, CHEESE, ETC.

Why has butter less flavor and color in winter than in summer? It is because the aroma, coloring principle, and flavors, are carried off from the green grass by heat, and the water of evaporation while the hay is being made.

All grains, flowering plants, and odorous vegetables, generally have certain valuable properties, aroma, flavors, or essential oils, ethereal in their nature, which are taken away by winds and the atmosphere, dried up by the temperature and the sunshine, or appropriated by the seed, if the plant be allowed to fully ripen.

What is more agreeable to the senses, than the gentle zephyrs of the ever welcome spring, wafting sweet aromas from fields redolent with honeysuckle and sweet clovers? How eager, with keenly awakened sense, we sniff the fragrant blossoms of the thickly budding orchard.

A whiff of the honey laden air, from the lee side of the buckwheat field in full blossom, is a fragrant joy never to be forgotten : while we inhale with ever renewed delight, the early summer breezes coming to us laden with the balmy perfume of sweet scented grasses with the wild violet, and delicate anemonie strewn amongst the tender blades.

These fine aromas, and pleasing scents, are not mere ethereal properties cast upon the air and having no value upon the nutritive character of the plant, but are indeed material qualities, having a marked effect on the appetite of the cow, and upon the palatable flavor of her products, and therefore desirable to be retained. Yes; it is *just these properties*, so valuable to the milk, which the parching sun distils from the dewy grass, and sips from the budding plant, only to be absorbed and borne away by the floating atmosphere ; but they should be retained, and made to impart their fragrance to the butter and the luscious cream.

In summer, too, when the cow can get such delicious food to eat, when she can crop the sweet green grasses, the succulent corn, and gather in the honeyed juices of the clover, is it any wonder that the milk and the butter should have a fine, agreeable odor and delicate flavor? For this tender, summer food is easily digested, and converted into protein and fat, supplying the nourishing casein of the cheese, and the delicious olein and margarine of the butter.

But in winter all these conditions are reversed, the most valuable properties of the summer pasturage have disappeared, the hay has lost the greater part of its flavor in drying, the nutritive juices of vegetable growths have been mainly absorbed in the reproducing seeds, consequently the butter and the cream, are pale, inodorous, and tasteless.

But can the winter butter be made to rival its more richly endowed predecessor of the summer? This is the question now disturbing the minds of the progressive dairymen and farmers. I believe the problem has received its solution in Ensilage; for the volatile properties of the bud and the flower are mostly saved and preserved by encelling the plant while in the green and aromatic state; the honeyed blossom of the clover can be made to retain its nectar in the milk, and the sweetest scent of the new mown hay, can be made to impart its fragrance to the butter. The encellated food, indeed, retains nearly all the flavoring properties of the fresh, green food.

It is well known to every farmer, that during spring and early summer pasturage, onions, leeks, garlics, cives, bitter mints, etc., impart unpleasant tastes and qualities to milk, and that their flavors are retained in the butter. There are also other unpleasant grasses occasionally, and accidentally licked up by the herb-cropping cow, greatly to the detriment of the milk and butter, which, indeed, are often times so injured by these flavors as to be rendered quite unpalatable.

Bitter and unpleasant flavors can be imparted *at will*, to the lacteal products of the cow. This may be easily proved, by feeding a milch cow with a mess of turnips, or plants of unpleasant odors.

I ask, if *disagreeable* tastes can thus be imparted to milk and butter, through feed to the cow, why cannot *agreeable* flavors and *pleasant* aromas likewise be imparted, and retained at will?

COLOR.

The color of milk and butter is effected by the various coloring matters in the herbs and grasses which the animal eats. The staining juices of many plants, and the various tinted chlorophyls, impart their hues to the milk; the green coloring principle of the grass, the yellow tint from the carrot, and the juices of other roots, all have their modifying effect upon the color of the milk and butter. The chlorophyl, which is the coloring principle of the green leaves and grasses, is much changed by drying, but in a great part is preserved by encelling.

Milk is of various hue and quality from different cows, even when fed upon the same food. This can be accounted for only by the peculiarity of the cow's nature, by her special physical character, and sometimes even by her own color, as the yellow, the white, and the black cow, are all liable to give a somewhat different shade of milk.

It is said, too, that being fed on carrots will color the cow's milk and the butter made from it. The food in winter differing so greatly from the food in summer, has a marked effect upon the milk, much of the coloring matter having been bleached out and dried away. The aroma, also, depends very much upon the kind of feed.

These, then, are the causes effecting the color, flavor and aroma of the various dairy products of the cow.

Perhaps the time is not far distant, when these coloring principles will be fed to the cow, in her winter food, and instead of putting annatto artificially into our butter in its making, we will supply our placid bovine with the means to color the butter herself.

The plants of summer certainly contain the coloring principle, and science will yet teach us how to retain it.

QUALITY AND QUANTITY.

That cows greatly vary both in the quantity and quality of their milk and butter, is a fact too well known to need discussion; but just why there should be such wide variations in these respects is still, to a great extent an open question.

Doubtless much depends upon constitution and physiological peculiarities.

Speaking in a general way, milk has always similar qualities, although there are certain variations in the composition and character of milk, yet they are limited in degree.

Whatever may be its source, milk contains in its natural state a nitrogenous element, as casein and carbonaceous elements, as fats and sugar.

The milk of cows, however, is not of *uniform* quality. Certain kinds as the Alderny, give a very large proportion of butter and but little milk, others give a large proportion of casein, and others an unusual proportion of water.

The kind and *quantity* of food, also greatly influence the quality of milk, so that with good provender and plenty of fresh, sweet grass, the milk is relatively richer in solids, and abounds in fats.

The effect of insufficient food on the quantity and constituents of milk was proved by Dr. De Caisne during the late siege of Paris. In his inquiries upon the effect produced by milk on children, he found that with much milk children thrived ; with little or bad milk, children had diarrhoea ; with scarcely any milk, children died. A well fed cow, then, means healthy children.

Now, while I firmly believe that ensilage combines and unites all the important qualities of a nourishing food for the cow, yet, I would caution the farmer who is experimenting with ensilage, against the ready error of ascribing to the encellated food, the variations and deficiencies which exist rather in the nature of the cow and in other definable causes ; for some cows differ in their reproductive and producing powers, their vital organs are more active and more powerful, they differ in their physiological structure, they are, in fact, differently constituted ; and just as one animal differs from another in any, or all of these natural conditions, so will their products vary.

Ensilage is rich in starchy compounds, and while there

is nothing lacking in it necessary to sustain life and support the *non-producing* animal in good condition, yet to the milch cow, for the enriching of her milk, it would be well, in addition to the ensilage to supply extra nitrogenous elements, such as bran, shorts, cotton seed, oil meals, &c. in proportion to the quantity of milk she gives.

WHAT IS BRAN?

Are farmers aware, that bran, as produced in our improved and modern mills is entirely different from the bran that was produced and fed to animals twenty years ago? While our bran formerly contained much of the gluten of the wheat which was highly nutritive, a great deal of the bran of the present day is but the mere external skin (or epocarp) of the wheat and contains *no nourishment whatever.*

To supply this gluten poor corn meal is now frequently mixed with the naked skins or epocarps of the wheat, giving the mixture the appearance of bran, so that what is now called *bran* is often but little more than adulterated meal.

This, as well as some other points that might be named are important, for the milk producing farmer who is feeding ensilage to consider, that ensilage may receive its due credit, and not be rejected or condemned for lacking qualities properly belonging to other foods.

The only way to judge ensilage fairly is to compare its results with the results of other food under like conditions.

Good feed to a healthy cow *will* produce good milk, and is more profitable for the farmer.

FERMENTATION.

Fermentation is the beginning of digestion. Flour taken into the mouth and mingled with the saliva which acts as a ferment, is changed into glucose, and like grape sugar is uncrystallizable.

Starch can be changed into glucose by being acted upon by various acids.

Some persons, in their first experiments with Silos and Ensilage, on opening the silo have found an alcoholic odor, and the ensilage somewhat sour. This is the *normal condition* of the ensilage, for I have never seen any that did not have this odor and slight acidity.

But this condition is no disadvantage to the ensilage; on the contrary it is that which imparts the agreeable flavor so much relished by the animal. Furthermore, the acids found in all ensilage assist in preparing the starch of the ensilage for being changed into glucose in the stomach of the animal.

They have also found portions of the ensilage mouldy, musty and rotten.

This condition is only found on, or near the top, and is owing solely to the air having been allowed to get in upon the surface where the action of the spore of *Penicillium* (always floating in the air with other fungi) produces *surface* mold. This should be remedied by firmly securing the spaces between, and at the ends of the planks.

As the atmosphere is one-fifth oxygen, and four-fifths nitrogen; when the ensilage is trodden down and the air forced out (excepting what remains in the interstices of the fodder) only a small quantity of oxygen can be left in it to excite fermentation or support spore life of the infusoria, which is always present when sugar is changed into alcohol; and we are taught by alcoholic examinations made by Pasteur and others, that oxygen in chemical combination with hydrogen and carbon becomes separated from the sugar, and goes to support the organisms which attend the work of fermentation with the change of the sugar, into alcohol.

As this ferment is connected with the work of living germs or spores (which have received the name of *Torula*,) and these spores live upon either the sugar or glucose contained in the liquid of grape juice, apple juice and other vegetable fluids, when all the sugar is broken up and changed into alcohol and carbonic acid gas, the spores must die by starvation for the want of more sugar, and thus *saccharine* fermentation will cease, when another fermentation is likely to be set up accompanied by another spore (Mycoderma aceti) which will oxydize the alcohol into acetic acid or weak vinegar.

Therefore, the amount of acetic acid formed, and of carbonic acid given off, can only be in proportion to the quantity of sugar, or alcohol that is oxydized, and to the quantity of oxygen consumed. The air that is retained within the interstices of the fodder, and which is not pressed out by packing and weighting of the ensilage, supplies the oxygen for oxydizing a portion of the existing sugar in the ensilage into alcohol and acetic acid.

The pickle is now perfected, and no further change can take place.

HEAT

A Silo filled with cut forage or corn stalks, will become heated in proportion to the quantity of sugar converted into alcohol and acetic acid. This can be easily proved by boring a hole through one of the central planks, thrusting down a rod or bar, and then lowering a thermometer into the hole, letting it remain a few minutes and then examining it. Carefully plug the hole afterwards. The Ensilage, will not rot, as the acetic and carbonic acids produced by the fermentation, pickle and preserve it. The acetic acid can undergo no change, unless while dissolved in the juices and other organic substances of the ensilage it becomes exposed to the atmosphere; and the carbonic acid, fumes of alcohol and aldc-hyde can only escape when the ensilage is disturbed.

Now such a condition cannot take place until after the Ensilage is taken from the silo and exposed for many hours in the open air; it will then, like all other organic substances, become mouldy, and decay, but it does not become injured under 24 hours of such exposure in ordinary weather.

I have made many experiments in preserving vegetable substances in pure carbonic acid gas, likewise, with pure nitrogen gas, also with pure oxygen, and have found that no fermentation in vegetable substances will take place if all the materials have been purified from living germs; proving to my mind conclusively that all fermentations are associated with some form of living organism.

I have preserved many vegetable substances by carbonic acid alone, and propose to build a cheap experimental Silo this season, in which I shall cause the Ensilage to be immersed in, and saturated with, carbonic acid gas as fast as it drops into the silo. I shall construct this silo as follows : Dig a round pit in the ground some 10 or 12 feet deep, in size, as circumstances may demand. Then plaster up the sides with good water-lime cement and gravel, in a manner similar to Fargo's rain water cisterns, using a heavy coat of cement to give strength to the walls, so they will stand the pressure of the surrounding banks of earth. At the lower part of this cell (or Silo) I shall construct a perforated floor, or diaphragm, raised a few inches from the bottom of the pit ; this bottom of the pit being made with a central hollow, or depression, in which a leaden basin or suitable vessel may be placed. Against the wall of the silo I shall arrange a tube passing under the flooring to the central basin, for the purpose of pouring in acids when required. Into this leaden basin (or other vessel) which is set into the hollow of the pit, I shall place a quantity of carbonate of lime, or other carbonates, which may be decomposed by causing mineral acids to run down the tube and come in contact with the carbonates, which will then unite with their bases and liberate their carbonic acid, which, ascending through the perforated diaphragm, will permeate the ensilage and destroy the spore germs and other vital organisms, which assist in and facilitate fermentation, thus preventing fermentation and other chemical changes of the ensilage. The silo might be partially filled with the gas before the chopped material is thrown in, or again, this tube may be sunk on the outside of the silo, or built into the wall with it as a permanent fixture, entering the space between the bottom of the pit and the perforated flooring through the lower part of the silo wall. By connecting this tube with a gas generator, or reservoir, set up in any convenient place, gasses can be forced down by pressure.

Carbonic acid or other gases may be forced down through this tube for the purpose of expelling the atmospheric air, and thus arresting any fermentation, mould, or other fungoid growth that may take place in free oxygen.

Food and fodder that I have prepared and submerged in carbonic acid, has an agreeable relish, and is readily eaten by animals.

The method of introducing the gas by the same pressure under which it is produced, is the most convenient, as it can thus be generated and regulated, and let into the silo in small or large quantities, as the silo is being filled with the chopped food.

The gas being heavy, it will remain at the bottom of the silo beneath the atmosphere, and thus its flow can be so regulated that during the filling of the silo the quantity of gas will be equal to the quantity of ensilage at any one time in the silo, and will thus submerge it. A silo made in this way must not be entered when opened without first testing, by means of a burning candle, the amount of carbonic acid present. It might otherwise prove fatal, as has happened in entering an exhausted beer vat. In my silo I shall arrange an apparatus by which the gas can be allowed to escape through a valve at the bottom before opening it. For several years past I have preserved fruits and vegetable substances by submerging them in an atmosphere of carbonic acid, and have found that they can be completely preserved in both flavor and color for one or two years, or until the cans were opened for use. I propose now to make this trial on a larger scale, and should it be as successful as with the small apparatus which I now have in my laboratory I shall report accordingly in an appendix to this pamphlet.

HEAT.

As all vital action is accompanied by heat, so all fermentations in organic bodies such as rot, mould and decay are chemical effects attended with vital action or, germs and spore life, and are accompanied with more of less heat as in the fermenting grape or apple juice while they are being changed into wine or cider ; this heat, increasing with the extent and rapidity of oxydization and the intensity of vital action.

ACIDS.

Acids are used in changing corn and potato starch into glucose and grape sugar. Consequently the ensilage in this state of mild fermentation from which the atmosphere is excluded, is in a condition when taken into the stomach of the cow, to have its starch at once acted upon by its acids and immediately converted into glucose and other compounds necessary to sustain and support the life of the animal and to enrich her milk.

We find the desire for a certain amount of acidity in food to be a universal one with man, and what is known as a *sub-acid diet* is the most wholesome, as a uniform sweetness palls upon the taste and cloys the apetite. The New Englander uses vinegar on his baked beans and boiled Cabbage ; the German on his *Sauer Kraut*; the Frenchman on his salad, and the Chinaman on his chow-chow.

Then why not quicken the appetite of the good Kine that yields us so many luxuries and fills our dairy with cheese and butter and bovine nectars? Give our faithful Ruminant a condiment by means of her encellated fodder. Farmers need have no fears that any bad qualities will be found in the milk, for this acid but excites the flow of saliva, and assists in changing the starch of the ensilage into the sugar which appears in the milk.

VINEGAR IN ENSILAGE.

The amount of vinegar in the ensilage is really but small compared with the amount of fodder. Green cornstalks without the leaves or tops, contain about 13 per cent of sugar. This sugar changed into alcohol could not produce more than six per cent of alcohol; and should this alcohol become changed into acid (vinegar) there could not be more than two per cent of acid in the ensilage, as a portion of the alcohol is evaporated before the acid fermentation takes place; so the quantity of acid in the silo would only be about equal to the proportion of vinegar that we use on our table vegetables.

Vinegar is an anti-septic. It preserves animal tissues and vegetable substances when they are excluded from the air, and it kills *Bacteria*, or spores that feed on animal putrefaction.

THE NUTRITIVE STATE OF GRASS.

It has long been known, and many wise farmers have acted upon the knowledge, that grass contains a greater amount of nutrition when in blossom, than at any time before or after.

Corn is a mammoth grass, and like grass should be cut

when in full bloom, as it then contains the greatest amount of sugar and nourishment in a soluble condition. All vegetable fibres become woody when the seeds, berries or fruits have ripened, and should therefore be preserved before this ripening period, when the elements that enter into their composition are in the most abundant and healthy condition, are easily digested, readily assimilated, and most favorably adapted for sustaining life.

Hogs and cattle will fatten more readily upon green corn in the milk, than they will upon the dried stalks and kernel, as then the entire elements of the food, being in a soluble state, are easy to be absorbed and assimilated by the tissues of the body.

Corn Oil is a carbo-hydrate. Hence the meal of corn and also of cotton seed is food for fattening animals, and likewise furnishes the buttery qualities of milk. I should therefore advise that they be made use of in feeding ensilage to milch cows.

Cotton Seed Oil not only has excellent milk producing properties, but the manure from it is highly valuable as a nitrogenous fertilizer, and, as a manure, is considered as valuable as before it went into the animal. Corn lacks this property.

Corn is so prolific that a given extent of land, cultivated with it, will yield a greater quantity of nutriment than by any other plant of the temperate zone fitted for animals and man, and is by far the cheapest food in all portions of the globe where it is grown.

Rye has been recommended as an addition to corn

ensilage, and it is a valuable one, for the rye being rich in nitrogen and magnesian phosphates it supplies to the corn ensilage the necessary elements for the production of rich milk, and gives to the food those qualities which are supplied by sprouts, and it may thus be found unnecessary to feed so much bran when the green rye is mixed with ensilage. Soil well manured ought to yield 10 tons of green rye per acre. This can be chopped and encelled alone or with the green corn stalks.

Clover and *Rye* are two nitrogenous producing plants which may be advantageously mixed with the corn ensilage, and I would advise in not more than one-fourth proportion to the corn.

The Cow Pea (or bean), being rich in nitrogenous elements, will also be found a valuable addition to the ensilage, in limited quantities, especially where clover or rye can not to be conveniently grown.

THE VALUE OF ENSILAGE AS A FERTILIZER.

It is useless to suppose that the earth is self-productive; that it can of itself reproduce its own elements; therefore, what the living principle organizes and takes away from it, must be restored or the soil will become non-productive, as the sterile and long neglected farms of our eastern states are every day telling us with all the pleading eloquence of helpless poverty.

The farmer who feeds his corn as ensilage and returns it to the soil as manure is replacing to the land all that has been taken from it, with the exception of the phosphates of potash, magnesia, lime, and soda, which are retained in the milk and disposed of in the sale of beef,

milk, butter, and cheese. And, in order that his farm
may be kept in a reproducing condition, these last named
elements must be restored to the soil in just proportion to the quantity of those elements that have been
disposed of in sending from the farm the milk, beef,
veal, mutton &c, or else, sooner or later, he will
find that his soil will have become exhausted of these
productive elements, and the true wealth of his farm
has been gradually disposed of through the sale of its
products; for it is idle to suppose that the earth can
yield to plants, elements which its soil no longer contains. It is easy to be seen, therefore, that if he is de-
sirous of continuously raising large crops of fodder, he
will be obliged to supply more manure than is yielded
by his ensilage fed stock ; and hence some of his fertilizers must be sought from other sources, as ground bones,
which are phosphate of Lime, and which enter largely
into the composition of milk, flesh, bone, and nerve; and
must be in them in order to perfect their normal qualities.

Fossil Organic Phosphates are now provided in large
quantities from the fossil beds of South Carolina; and
mineral phosphates such as *Apatite* (crystal phosphate of
lime) are found in large quantities in Canada and Nor-
way, and with other forms of phosphates, in other portions of the globe ; all of which are now necessarily
becoming staple articles of commerce.

POTASH.

Again, as Potash is consumed or taken in large quantities from the earth by the corn, the soil will eventually

become exhausted, and unless the potash be given back
to the soil again, the crop will finally become limited in
its growth, and the potash must therefore be supplied
by artificial means if a continuous crop be desired.

Within the last few years inexhaustible sources of
potash salts have been discovered in the geological for-
mations of Europe, and are now supplying the world.

Wood Ashes are very valuable to the farmer, as they
contain potash, phosphorous and other inorganic substances which must be contained in every soil that yields
large crops of life sustaining food.

As nitrogen is contained in large quantities in all
organic products, a nitrogenous element must exist in
the manure which is used to replenish the soil, as in all
cases we must in some way return to the earth the equiv-
alent of what we have taken from it.

Now the use of ensilage, as fodder, restores to the soil
as manure all those elements *not sold off*, which are necessary to the reproduction of farm products, and our farmers will therefore find in ensilage the long demanded
and much wished for fertilizer. All these principles of
artificial fertilizers enumerated above, refer more aptly to
the old and nearly worn out farms of the Eastern and
Atlantic States, but will sometime be applicable to the
prolific soils of the west, as a continuous cropping must
finally exhaust even the richest and most productive soil.

In a treatise on Agricultural Chemistry intended to
follow this pamphlet, I hope to be able to set forth these
principles more fully. I shall state them in a simple
and comprehensive manner, within the ready under-
standing of every reader.

A word to our southern farmers.

I would like here to suggest an experiment to southern farmers and cotton producers.

It is well known, for chemical examinations have proved, that the *green* cotton plant, while in its tender state, is rich in nutritive qualities, which under the present system of cultivation are usually lost by the stalks being allowed to dry and decompose in the field after the cotton has been gathered. As for some time after the cotton is picked the stalk remains green and juicy, the plant could be cut and added to the ensilage, and as it contains all or many of the life sustaining qualities it might become one of his most valuable additions to encelled fodder.

Selling the cotton seed: When the southern farmer sells his cotton seed and allows the stalk to dry up, he is carrying away nearly all the nitrogenous principle of the plant, and much of the more valuable portions of both the organic and inorganic elements of the plant. But by encelling his green cotton stalk he can make his cotton plant fodder as valuable, per ton, for his stock as the average crop of hay is to the northern farmer.

CLIMATE.

Considerations of climate need not deter the southern farmer from trying silos and ensilage, for if the silos are successful in France, Spain and Italy, why should they not be in our middle and southern States?

Well constructed, and well secured silos, will preserve food in the southern States, if the fodder be taken from the field fresh, cut fine and immediately compressed in the silo and the air excluded from it. As it is warmer in the south, no time should be lost in securing it from the warm air, as light and increase of temperature facilitate chemical changes in animal and vegetable substances.

HONOR TO WHOM HONOR IS DUE.

First experiments in Ensilage. M. Augustus Goffart is the man whom farmers in all parts of the world must honor as being the first to make successful experiments for preserving green food for cattle without drying or haying. He began his experiments about 30 years ago. He has done more to improve cattle feeding and to save labor in tending stock than any other man, and his name ought to be perpetuated in the grateful memory of all men honoring worthy deeds. He has not only enabled man to store food for animals in its original juices, and thus to preserve all its nutritive qualities, but to preserve them in a more palatable condition, and with an important preservation of valuable qualities heretofore lost by drying and haying.

He has shown how an immense saving of space is gained by dispensing with expensive barn room. He has also done away with a great deal of arduous labor ; the work of curing hay under the sun's broiling heat is now swept away, for the *green* grass is put into the Silo ; and not only the grasses, but the smaller grain bearing cereals can be encelled with the maize or stalks of corn.

WHAT DOES ALL THIS REVOLUTION BRING ABOUT ?

The farmer who stores his fodder in silos has so much less to fear from lightning or whirlwinds; from the torch of the incendiary, the careless match, or neglected pipe of the lodging tramp.

He saves insurance on costly barns and stores the nourishment of the bulky ton of hay in the space of a few cubic feet, where fire cannot consume, nor winds destroy. This revolution means more healthful milk, butter, cheese and all the products of the dairy raised on our farms in increased quantities, and at less price than from the distant west.

And the young farmer with only a few modest roods of ground, can now outrival the broad acres of his ancestral predecessor; and even the mechanic with his little village lot can with equal readiness supply his family with the luxury of pure milk and sweet odorous butter.

This revolution, too, means that by the use of encelled food there can be raised near our large cities and sea port towns, beef, mutton, and pork in quantities to well compete with the interior States; hides and tallow can be found at home instead of being imported so largely from abroad; and all this class of productions, horses, sheep, wool, hides and tallow can be so cheapened that the east will no longer be dependent on the west; for the small neglected farms of our eastern States can now be made to vie with the large crops raised in the rich and fertile prairies of the far west. Freights would consequently be reduced, and thus the immediate home products of every farmer could each and all contribute their quota toward pulling down monopolies, and putting a stop to preferred freights. In fact, the amount of stock, and stock products the eastern States could thus raise at home, would be so great that they could be made almost wholly independent of the west, and the staple articles most in demand could be supplied from our own neighboring lands.

This revolution of M. Goffart means a liberal supply of sweet wholesome fodder to animals, instead of feeding them on sour slops from whiskey stills, hotel swill, and other refuse food which are better for fertilizing the soil, than for producing milk unwholesome for babes and invalids ; milk quivering with *Vibric* and *Bacteria* engendering disease and death.

This revolution once fairly inaugurated there will be in our towns and cities less *Cholera Infantum* and other complaints, which now carry off their thousands every summer ; there will be fewer new homes made desolate ; fewer weeping mothers with hearts sadly turning to cemeteries closely strewn, with little graves.

M. Goffart has done more for the agricultural interests of the world than any other man, living or dead. He has done for the agricultural interests of this country what Lafayette did for our national liberties, and the blessings his experiments have conferred upon our national productions, in their effects upon humanity at large, find their fitting parallel only in the benefits conferred upon our nation by the sword of his illustrious countryman; and should he ever honor this country with a visit he will find as our patriots received Lafayette, so will the hearts of the farmers of our nation receive him. All our agricultural societies will welcome him with highest honors, and every tiller of the earth will gratefully preserve his name in everlasting memory.

Where honors are due there let honors rest.